Geometric Patterns

from
Churches & Cathedrals

Robert Field

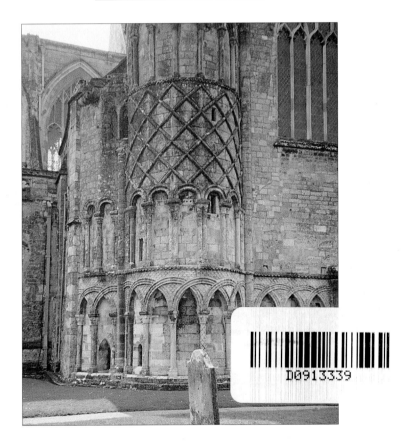

Tarquin Publications

Wenlock Priory, Shropshire

All photographs and illustrations in this book are by the author with the exception of the following: *page 9 Exeter Cathedral - Peter Ostrolenk; page 10 Kings College Chapel, Cambridge - Jean Aston; page 44 - Bristol Cathedral - Dr. Pat Witts.* The author wishes to thank them for granting permission for their photographs to be included.

© 1996:   Robert Field
I.S.B.N.:   1 899618 13 9
Design: Magdalen Bear
Printing: Ancient House Press, Ipswich

Tarquin Publications
Stradbroke
Diss
Norfolk IP21 5JP
England

Gloria in Excelsis

Man's inventiveness is a constant source of wonder and nowhere more so than in ecclesiastical buildings. Design, harmony, balance and symmetry are all words which spring to mind and the purpose of this book is to draw attention to churches and cathedrals as a rich source of inventive geometric patterns.

The artists and architects used the feelings of harmony and symmetry suggested by such patterns as a statement of their religious beliefs and the results of this artistic expression can be seen in the structure of the building, in its doorways, its windows and flooring, as well as in the decoration of its fixtures and fittings.

Many churches and cathedrals have survived over hundreds of years and many were restored and extended in the Victorian era. New churches were built then and it was a time of great confidence, invention and the development of new materials. British architects of this period looked to the past for the inspiration for their own needs and were prepared to use and to adapt designs from other countries. Therefore, churches built in the last century provide us with perhaps the greatest opportunity to find interesting ideas.

These patterns and designs can be appreciated for their own sakes and the collecting of them can provide hours of enjoyment. Once observed and noted, the designs can be developed and used in other work. For instance, in drawing and painting, marquetry, needlework, collage, patchwork and in fabric designs.

This book does not aim to be an exhaustive study but is intended to direct your eyes to where to look for patterns and designs in churches and cathedrals and what you might find. I hope you derive as much pleasure from them as I have done in searching out subjects to photograph for this book. Happy hunting.

Robert Field

Winchester Cathedral

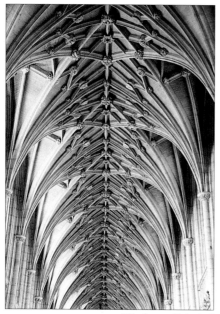

Winchester Cathedral

Winchester Cathedral

The first pattern that attracts the eye on entering a church or cathedral is the structure of the building itself. Look at the shape of the arches and windows, the repetition of these shapes along the length of the building and the pattern of the vaulting on the ceiling. The detail is usually unique to each building, but there are recognisable styles associated with the period when it was built. It is interesting to learn to recognise those styles and their approximate dates and also to be able to see reinterpretations of those styles in Victorian and more recent buildings.

Salisbury Cathedral

Christchurch Priory, Dorset

Sherborne Abbey, Dorset

Architectural Styles

When looking at the structure of ecclesiastical buildings, there are four main styles of Gothic architecture and they each have distinctive features which help with identification. The styles are called Norman (c. 1066 - 1200), Early English (c. 1200 - 1300), Decorated (c 1300 - 1400) and Perpendicular (c. 1400 - 1500).

Norman (c. 1066 - 1200)

The strongest features of the Norman style to recognise are the rounded arches and the thickness of the masonry.

Doorways consisted of a series of arches stepping backwards and decreasing in size. They were generally decorated in a distinctive geometrical way using triangles, squares, lozenges, zig-zags and circles, all in low relief. Such patterns were also used on columns, the capitals of the columns and on wall surfaces.

Such doorways were generally not very wide as the weight of the massive walls above had to be supported. For the same reason, the openings for windows were kept quite small and narrow.

As the building techniques improved during this period, it became possible for doorways and windows to be larger and more light could penetrate to the interior. Then, towards the end of the era, the stronger pointed arch was invented and architects were able to let even more light into the building and span wider areas with vaulting.

St. Nicholas, Barfreston, Kent

Colchester Abbey, Essex

6

St. Mary, Portchester, Hampshire

St. Mary, Portchester, Hampshire

Here are some of the geometric patterns to be found in Norman architecture.

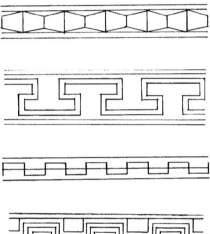

Canterbury Cathedral

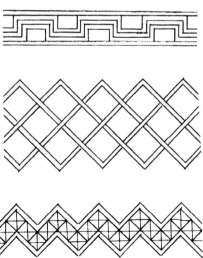

Romsey Abbey, Hampshire   Salisbury Cathedral

Early English (c. 1200 - 1300)
With the development of the pointed arch, walls could be much thinner and yet still
support the weight of the roof. The huge thick columns of Norman times became
unnecessary and were replaced with groups of thin columns around a central shaft.
Narrower and taller windows give a much greater sense of height and lightness to the
buildings. With buttresses between the windows absorbing the horizontal thrust of the
roof and stone cutting becoming more accurate, the walls could be made thinner still.
Stone carvers also developed greater skill with their chisels and no longer carved just a
surface decoration in low relief, but worked in deeply undercut high relief. Designs
based on naturalistic plant forms became popular and geometric patterns less common.

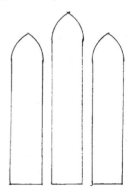

Milton Abbey, Dorset

Decorated (c. 1300 - 1400)

As its name suggests, the buildings of this period show a much greater degree of ornament than earlier. Look particularly at the windows. It is the tracery within them which gives this style its distinctive flavour. As the window openings became larger, the arches were partly filled by a wonderful curvilinear tracery, often based on the ogee. Its shape is illustrated below and is developed from touching circles. Columns became taller and more slender and the roof vaulting more decorative.

Exeter Cathedral

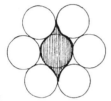

Canterbury Cathedral

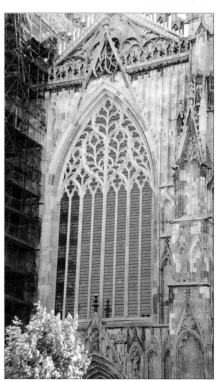

York Minster

Perpendicular (c. 1400 - 1500)

The wonderful fan vaulting and the huge windows are the most recognisable features of this style. Within the windows, the tracery is much simpler and more vertical than was the case during the exuberance of the earlier 'Decorated' period. Another distinctive feature is the four-centred arch which gives the tops of the windows and doorways a more flattened appearance.

Kings College Chapel, Cambridge

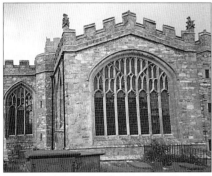

St. Beuno, Clynnog Fawr, Gwynedd

The Cloisters, Gloucester Cathedral

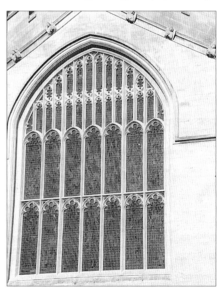

St. Mary the Virgin, Oxford

Post Gothic Style

The four main periods of architecture described so far are often referred to as Gothic. Strictly speaking, the word Gothic means the architecture of western Europe from the 12th to 16th centuries, characterised by pointed arches.

To continue to follow the development of patterns in the structure through the post Gothic era and into modern times in a book such as this is not terribly productive. That does not mean to say that there is anything lacking in the structures of Tudor, Jacobean, Georgian or the revival of classical styles with Inigo Jones and Christopher Wren. Simply that it is more interesting from a geometrical point of view in later churches and cathedrals to concentrate on the detail and the decoration. That is where the new and ingenious developments in geometrical patterns are to be found.

Later periods developed their own styles and made use of new materials, but also looked back to the Gothic period for inspiration and reinterpretation. This was especially true of the Victorians. Their preferred material was brick and as the photographs below show, they made good use of its decorative properties. Such decoration had been used on brick buildings since Tudor times but the Victorians had a greater range of colours at their disposal and were also more innovative with their patterns. Such designs can be seen especially well in Methodist chapels built during the century which followed their formal separation from the Church of England in 1791.

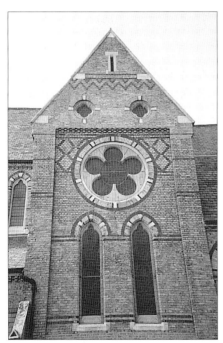

St Peter, Wapping, London

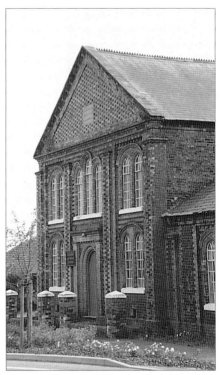

Methodist Chapel, Telford, Shropshire

Decorative Ironwork

As wrought iron became cheaper and more widely available, it was introduced more and more into ecclesiastical buildings. Look how it has been used for gates, grills, altar rails, hinges on wooden doors and covers for the font. Many motifs have been copied in a stylised way from nature and plant forms but developed into geometric patterns.

Salisbury Cathedral

St. Mary, Shackleford, Surrey

St. James, Kingston, Dorset

St. Mary the Virgin, Silchester, Hampshire

St. Mary, Easton, Hampshire

St. Thomas, Salisbury, Wiltshire

Victorian Gratings

As hot-water heating systems came to be installed in churches, the design of iron gratings to cover the pipes allowed great inventiveness. From a practical point of view, all that was needed to allow the heated air to circulate in the building.was a series of perforations. However, artists rose to the challenge and produced many geometric patterns of outstanding beauty and interest.

Patterns of intersecting circles were frequently used and some designs show obvious links back to Roman mosaics. Others show interlinking, typical of Celtic designs.

On this page and on the following six pages are photographs and drawings of a fine selection of interesting designs.

St. Eata, Atcham, Shropshire

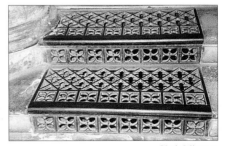

York Minster.

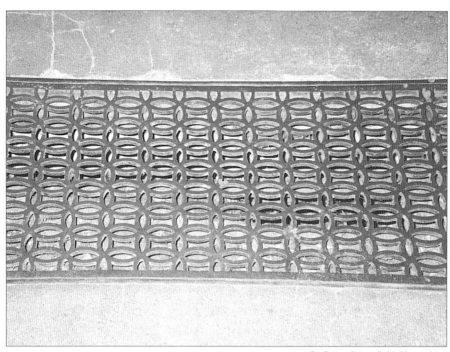

St. Peter, Petersfield, Hampshire

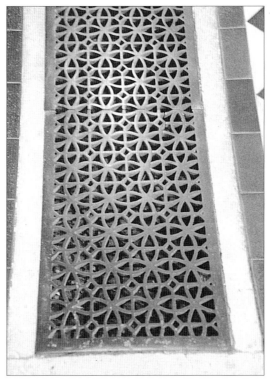

St. John the Baptist, Bere Regis, Dorset

St. Mary the Virgin, Silchester, Hampshire

The same basic design can be developed into a different pattern of overs and unders.

St. Matthew, Blackmoor, Hampshire

St. Peter, Hornblotton, Somerset

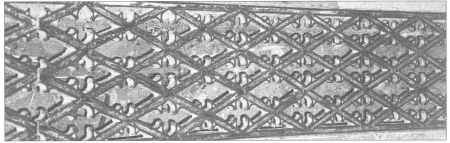

St. Andrew, Aldborough, Yorkshire

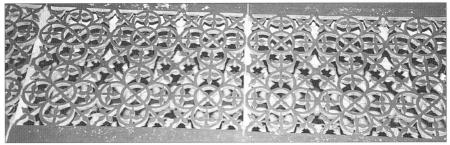

Wimbourne Minster, Dorset

York Minster

St. Andrew, Aldborough, Yorkshire

Leaded Lights

Until relatively recently, it was not possible to produce glass in large sheets. Windows therefore had to be made of small panes of glass set into and separated by strips of lead. The small panes could be of any shape and the glass could be coloured or plain and the tracery of these strips gave a strong geometric structure to the window. Artists and architects took advantage of it in two distinct ways.

In medieval times, coloured glass was used to make pictures which told the Bible stories, and the results can be seen in the wonderful stained glass windows that grace churches and cathedrals throughout Europe.

Cerne Abbey, Dorset

In later periods, many architects wanted to let more light into their buildings and so plain-glass leaded windows became more common. In such windows, the design of the network of lead strips becomes dominant and the emphasis is on the pure geometry of the pattern. In the pages which follow there are many examples of the inventiveness of the way leaded lights have been treated through the centuries.

St. Peter, Petersfield, Hampshire

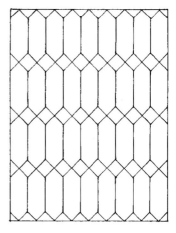

St. Mary, Morden, Dorset

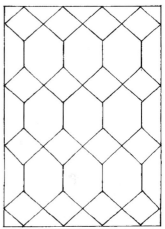

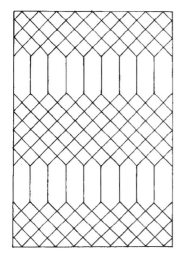

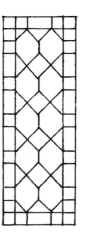

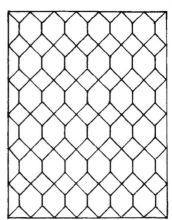

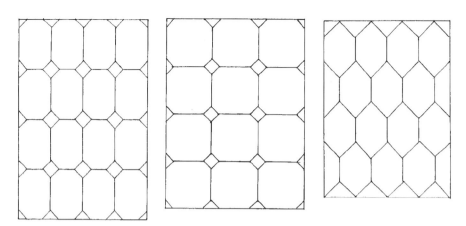

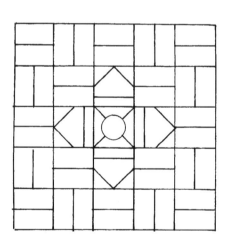

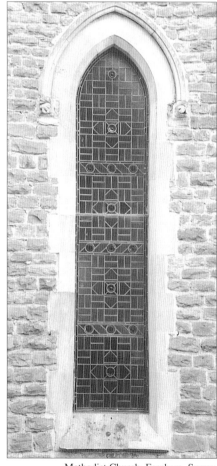

Methodist Church, Farnham, Surrey

St. James, Milton Abbas, Dorset

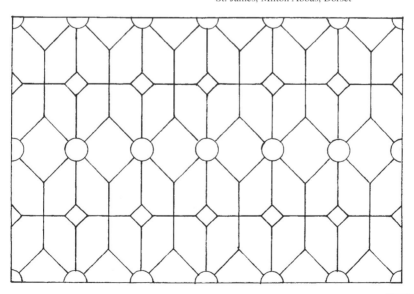

St. James, Milton Abbas, Dorset

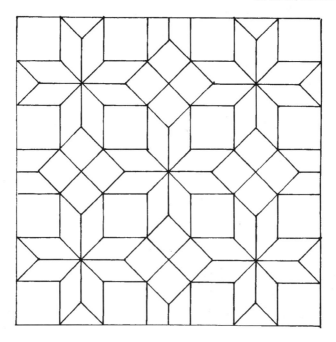

St. Mary Major, Ilchester, Somerset

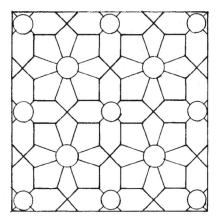

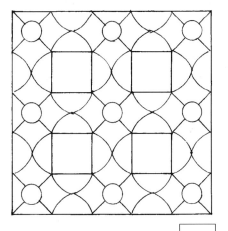

St. Mary Major, Ilchester, Somerset

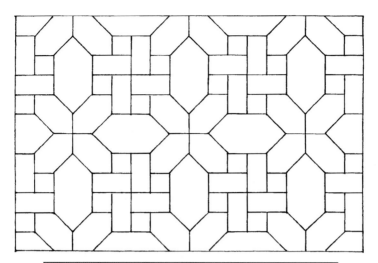

Holy Trinity, Guildford, Surrey

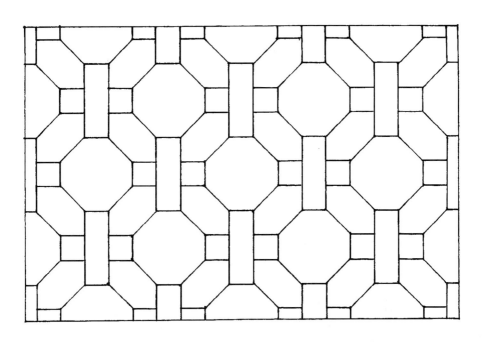

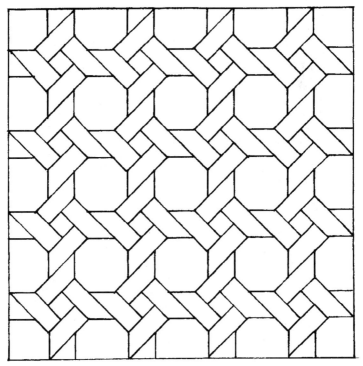

The Windows of Canterbury Cathedral

At the east end of Canterbury Cathedral there are 13th century windows which have iron armature supports called ferramenta. These supports create a strong geometrical design within which the pictorial stained glass is set. They make an interesting study of the variety of ways in which a window shape can be divided.

Canterbury Cathedral

Canterbury Cathedral

Canterbury Cathedral

Canterbury Cathedral

Canterbury Cathedral

29

Canterbury Cathedral

Canterbury Cathedral

Canterbury Cathedral

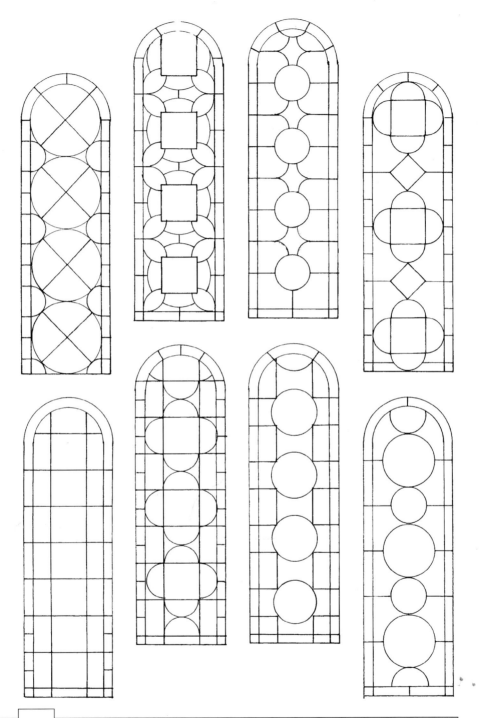

The organisation and design of stained glass windows often reveal a geometric scheme. This example has a strong design of squares and touching circles.

Canterbury Cathedral

Grisaille work is the name given when the windows have patterns painted on the glass in monochrome.

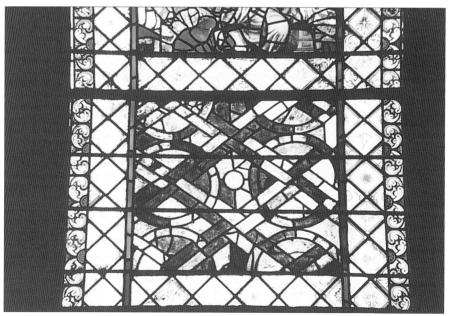

Salisbury Cathedral

Both of these patterns come from Salisbury Cathedral

Rose Windows

One of the glories of Gothic architecture is undoubtedly the Rose Window. Originally, it was a circular opening within a square but it developed into a symbol representing the wheel of life. Further symbolism was added, especially in France and those at Chartres and Notre Dame in Paris have complex geometric and iconographic schemes. If a visit is possible to either then it is well worth the services of a guide in order to understand its full significance and meaning.

San Francesco, Assisi, Italy

Santa Chiara, Assisi, Italy

St. Nicholas, Barfreston, Kent

Westminster Abbey

Notre Dame, Paris

Rouen Cathedral, France

Not all circular windows are Rose Windows and they are often to be found in Victorian churches. The designs here show many different and ingenious ways of treating a circular aperture.

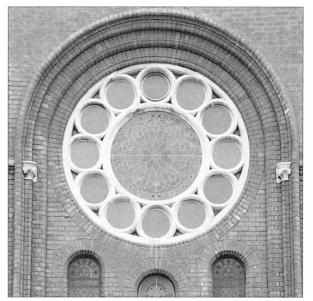

Baptist Church, Salisbury, Wiltshire

St. James, Kingston, Dorset.

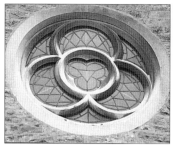

United Reform Church, Farnham

St. Mary, Swanage, Dorset

United Reform Church, Farnham

St. Mary, Swanage, Dorset

Cosmatesque Floors

In Italy during the 12th century, members of the Cosmati family were involved in the building of many churches. They favoured a particular style of flooring which made use of coloured marble cut into geometrical shapes. These shapes were laid in patterns which were derived from Roman mosaic, opus sectile and Byzantine designs. A particular recognisable characteristic is that the tiles curve around circles: This style of floor decoration became known as Cosmati work and many examples have survived in churches in Rome and elsewhere in Italy.

Salerno Cathedral, Italy

Salerno Cathedral, Italy

These designs all come from the church of Santa Maria in Cosmedin in Rome.

When Richard de Ware was made Abbot of Westminster in 1258, he went to Italy to be confirmed in the post by Pope Alexander IV. The ceremony took place at Anagni Cathedral which has a wonderful Cosmati floor. The Abbot was impressed and as a result brought back craftsmen and materials to England to lay a similar floor in Westminster Abbey. These pages show some of the patterns to be seen on that floor.

Some Victorian churches have marble and tile floors which were clearly inspired by mosaic, opus sectile and Cosmati designs.

Bristol Cathedral

St. Peter, Hornblotton, Shropshire

St. James, Weybridge, Surrey

St. James, Weybridge, Surrey

Byland Abbey, Yorkshire

Medieval Mosaic Tiles
In medieval times in Britain and elsewhere in Europe, the floors of abbeys, churches, monasteries and other important buildings were tiled. The first of the two main types of tiling was derived from Roman mosaic and opus sectile floors. The tile-makers manufactured large quantities of standard triangles, squares, hexagons, rectangles and other geometric shapes. They were then fired and glazed in different colours and laid in intricate patterns. Sometimes the inspiration for the patterns can be seen in floors elsewhere but the large circular patterns involving many different shapes were original and developed on the site. The best examples in Britain of this type of floor are to be found at the large monastic settlements in Yorkshire such as Byland Abbey. The following pages show diagrams of other designs which have been discovered.

Byland Abbey, Yorkshire

46

St. Cross, Winchester, Hampshire

The second type of tile utilised for medieval floors made use of a different technique. Each tile was made with red clay and, while still wet, an impression was made in its surface by pressing it with a wooden pattern block. This impression was filled with another lighter coloured clay to make a two colour design. It was then fired and glazed. The designs on individual tiles were not always complete in themselves but were used with others of the same or different design to create a larger pattern. They were grouped either in fours (2 x 2), nines (3 x 3) or sixteens (4 x 4). Often areas of a design were framed with rectangular tiles of a different colour or glaze. Not all the designs were geometric; some, like those produced at Chertsey Abbey, were pictorial. Often the patterns were based on ones from antiquity.

Winchester Cathedral

The illustrations on this page and the two following, show individual tiles together with patterns which can be made from them. Some are simply placed side by side, others need to be rotated in order to create the larger pattern.

Single Tile

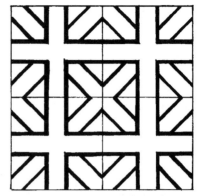

Pattern of Four Tiles

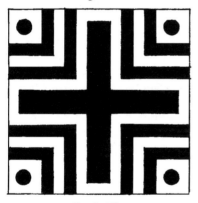

Single Tile

Pattern of Four Tiles

Single Tile

Pattern of Four Tiles

Single Tile

Pattern of Four Tiles

Single Tile

Pattern of Four Tiles

Single Tile

Pattern of Four Tiles

52

Single Tile

Pattern of Four Tiles

Single Tile

Pattern of Four Tiles

Single Tile

Pattern of Four Tiles

Patterns made of nine tiles require three different designs.

Four of these   Four of these   One of these

The pattern of nine tiles

Patterns of sixteen tiles are more complex and need four tiles each of four different designs. Two of these tile designs are mirror reflections of each other.

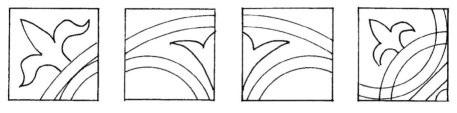

Four of each of these tiles

The pattern of sixteen tiles

Victorian Tiles

In Victorian times, many new churches were built and older ones renovated. Since many of the old floors were worn out, architects needed a supply of new flooring material; in particular, tiles. They approached the tile manufacturers of the day and asked them to make replacements. The new tiles were of course machine-made, but many of the designs and patterns were based on earlier hand-made medieval examples. In some cases they reproduced exact replicas of the tiles that they replaced. In other cases, the new tiles were designed by prominent artists such as Pugin.

This was also the era of the plain coloured tile. Accurately made and expertly laid, they explored in full the geometrical possibilities of the tiled floor. Many are still in use today.

Milton Abbey, Dorset

Milton Abbey, Dorset

Milton Abbey, Dorset

Holy Trinity, Leeds, Yorkshire

Holy Trinity, Leeds, Yorkshire

Designers took delight in creating elaborate patterns using very few different shapes and colours of tile. These examples show the kinds of effects that can be achieved.

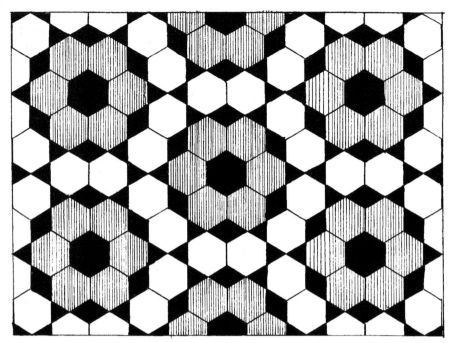

The pattern above requires only tiles which are hexagons and triangles. The one below also needs squares. Notice how the edge lengths have to match.

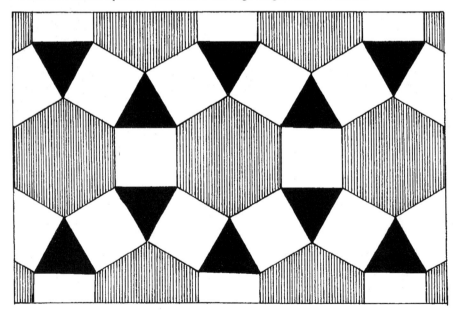

St. Andrew, Shifnal, Shropshire

St. Andrew, Shifnal, Shropshire

Chichester Town Hall, Sussex

Look out for tiling patterns like these wherever you go. If you carry a pad of squared graph paper with you, it is quick and easy to sketch the outlines of what you see.

Sketching in this way is a good technique for increasing your powers of observation and becoming more aware of the richness and variety of the geometrical designs and patterns which can be found in churches and cathedrals.

Both the examples on the right are from Bere Regis in Dorset.

St. John the Baptist, Bere Regis, Dorset

St. Peter, Leeds, Yorkshire

St. Bartholomew, Ranmore, Surrey

St. Nicholas, Moreton, Dorset

St. Nicholas, Abbotsbury, Dorset

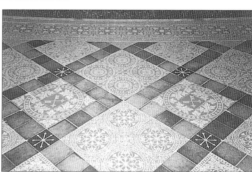

The Chapter House, York Minster

St. Andrew, Trent, Dorset

Let us finish with a few more examples of geometrical patterns in ecclesiastical buildings. They can be anywhere, inside or out. Look at panelling, screens, bench ends, railings, fonts, columns and memorials, always keeping geometry in mind.

You will not be disappointed.

St. Mary, Portchester, Hampshire

St. John the Baptist, Bere Regis, Dorset

St. James the Less, Pimlico, London

St. James, Milton Abbas, Dorset

St. Andrew, Shifnal, Shropshire